Pramod Singh (Ed.)

Salicylic Acid and Nutrient Regulation

Pramod Singh (Ed.)

Salicylic Acid and Nutrient Regulation

Sustainable Productivity

LAP LAMBERT Academic Publishing

Impressum / Imprint
Bibliografische Information der Deutschen Nationalbibliothek: Die Deutsche Nationalbibliothek verzeichnet diese Publikation in der Deutschen Nationalbibliografie; detaillierte bibliografische Daten sind im Internet über http://dnb.d-nb.de abrufbar.

Bibliographic information published by the Deutsche Nationalbibliothek: The Deutsche Nationalbibliothek lists this publication in the Deutsche Nationalbibliografie; detailed bibliographic data are available in the Internet at http://dnb.d-nb.de.

Coverbild / Cover image: www.ingimage.com

Verlag / Publisher:
LAP LAMBERT Academic Publishing
ist ein Imprint der / is a trademark of
OmniScriptum GmbH & Co. KG
Heinrich-Böcking-Str. 6-8, 66121 Saarbrücken, Deutschland / Germany
Email: info@lap-publishing.com

Herstellung: siehe letzte Seite /
Printed at: see last page
ISBN: 978-3-8473-4937-2

SALICYLIC ACID AND NUTRIENT REGULATION

Sustainable Productivity

Author:

Pramod Kumar Singh

Affiliation:

Udai Pratap Autonomous College, Varanasi-221002, INDIA.

Affiliated to Mahatma Gandhi Kashi Vidyapeeth University, Varanasi-221002, INDIA.

e-MAIL:

drpksingh2001@yahoo.com

Mobile:

91- 9415388189

CONTENTS:

1. Introduction

Plant growth is the ecologically and agriculturally important process of biomass accumulation and seed production are affected by a number of environmental factors. Inorganic ions also represent an integral part of the environmental cocktail that affects plant growth and productivity. Sustainable development is the key requirement for proper growth and yield of plant species under changing environment due to adverse human activities on the natural ecosystem. Non-scientific utilization of natural resources cause socio-economic imbalance worldwide. It is now understood that agriculture also can negatively affect the environment through over use of natural resources as inputs or through their use as a sink for waste and pollution. Such effects are called negative externalities because they impose costs that are not reflected in market prices (Pretty *et al.*, 2012). These environmental costs shift conclusions about which agricultural systems are the most efficient and suggest that alternative practices and systems that reduce negative externalities should be sought. Food security to the present population needs high adaptation ability of plant system exposed to adverse environmental conditions (Singh *et al.* 2011). The agricultural productivity attained during the 1980s has not been sustained during the 1990s and has posed a challenge for the researchers to shift the production function upward by improving the technology index.

Green revolution drove this production growth with new varieties, fertilizer inputs, water management and rural infrastructure most increases in food production were achieved on the same agricultural land. Agricultural growth needs higher production with soil fertility. These indicate that there has been a need of optimal production growth across the world. Continued population growth, rapidly changing consumption patterns, and the impacts of climate change and environmental degradation are driving the limited resources of food energy water and materials toward critical thresholds. Thus the challenge still remains substantial for agricultural production under adverse environmental conditions. Sustainable agricultural intensification is defined as producing more output from the same area of land while reducing the negative environmental impacts and at the same time increasing contributions to natural capital and the flow of environmental services (Pretty, 2008; Godfray et al., 2010). Role of phytohormones is very critical in growth and developments of plants (Rubio et al. 2009). A major goal of phytohormonal ecology is to understand the role of phytohormones in determining plant responses to various environmental challenges. This includes the strategy that each hormone plays in response to individual stress as well as the coordination of multiple stresses. However, complex defense signaling network in nature should join forces to place molecular mechanism of induced plant defenses in an ecological perspective. One particular challenge is to unite the knowledge of phytohormonal biochemistry with patterns of ecological correlation among stresses. One possible strategy to understand the interaction between stresses is hormonal and biochemical signaling which may influence the severity or the ability of plants to adopt against various stresses (Singh *et al.* 2011). Current biochemical and molecular models of phytohormonal signaling pathways as predominantly responsible for plant resistance to adverse environmental conditions may be utilized for the genetic identification of signaling components in plants. Salicylic acid, Jasmonic acid and Abscicic acid are recognized as key players in the regulation of the signaling pathways (Wang *et al.* 2007). But complete cross-talk signaling pathways and their receptors require to be elucidated. Salicylate pathways for signal transduction are the key biochemical response mechanism that can be triggered by various metabolic changes during growth and developments of plants.

Salicylic acid (SA) is a phenolic compound synthesized throughout the plant kingdom via the phenylpropanoid pathway (Metraux, 2002). Research efforts over the past decade have focused on this molecule to elucidate its many roles in plant physiology. Detailed evidence implicates SA in pathogenesis-related (PR) gene expression, systemic acquired resistance, and the hypersensitive response (Shah, 2003). SA also seems to be involved in responses to abiotic stresses, such as ozone (Rao and Davis 1999; Koch *et al.* 2000), salt and osmotic (Borsani *et al.*, 2001; Molina *et al.*, 2002), UV-B (Surplus *et al.*, 1998; Nawrath *et al.*, 2002), drought (Nemeth *et al.*, 2002; Munne-Bosch and Penuelas, 2003), and metal stress (Metwally *et al.*, 2003; Yang *et al.*, 2003; Pal *et al.*, 2005). However, the effects of salicylic acid on plant resistances to abiotic stress usually contradict with each other. The same pre-treatment with exogenous SA results in opposite responses in different plant species (Yang *et al.*, 2004). Even the same SA concentration promotes resistance to one kind of stress meanwhile decreases the resistance to another stress (Nemeth *et al.*, 2002). The fact that SA can exert different effects under various stress situations or with different species is not in contradiction but rather illustrates the fact that different stresses can either is dependent or independent on an SA pathway and that the molecule does not have the same effect on various species.

Salicylic acid (SA) is considered to be a potent plant hormone because of its diverse regulatory roles in plant metabolism and act as an endogenous plant growth regulator of phenolic nature that possesses an aromatic ring with a hydroxyl group or its functional derivative. SA is known to affect growth and bio-productivity in the crops (Hayat et al. 2010). It is also recognized that SA potentially generates a wide array of metabolic responses in plants which affects photosynthetic parameters and plant water relations (Fariduddin et al. 2003; Vicente and Plasencia 2011). The oxidative damage caused to the plants due to the increased generation of reactive oxygen species under stressful environment could be mitigated by the external application of SA (Hayat et al. 2010; Singh and Gautam, 2013). A large body of literature demonstrates the effect of SA seed soaking on plant growth and development. However, very little is known about the effect of SA applied to foliage on plant growth and development. Keeping in view the diverse physiological roles played by SA, the present concept was undertaken to improve our understanding of the effect of the various concentrations of SA applied as foliar spray on the growth, photosynthesis, nitrogen fixation and its assimilation in chickpea and to screen for the most effective concentration of SA. The content of any active protein (enzyme) represents a fine balance between its synthesis/activation and degradation/inactivation (Jain and Srivastava 1981). The concentration of SA might play an active role in such a regulation where the lower concentration favored an increase in the NR protein and higher quantity of SA decreased it. Moreover, the reason that seems to be most appropriate to explain the SA mediated elevation in the activity of NR is that it stabilizes the membrane structure and its fluidity which could have facilitated the increased uptake of nutrients including nitrate thereby increasing its content in the roots, which also acts as an inducer of NR (Campbell 1999). The increase in the content of nitrates and thereby activity of NR due to exogenous SA treatment under normal growth conditions was reported earlier (Hayat et al. 2005).

Carbonic anhydrase (CA) is the other most abundant soluble zinc containing enzyme in chloroplasts of C_3 plants, after Rubisco and facilitates the diffusion of CO_2 across the chloroplast membrane by catalyzing the hydration of dissolved CO_2 as it enters the most alkaline environment of stroma (Majeau and Coleman 1994) where CA catalyzes the reversible hydration of CO_2 and maintains a constant supply of Rubisco. The content of the enzyme and therefore, its activity is under the fine regulation at the level of transcription and/or translation (Okabe et al. 1980). However, either of these processes was favoured by the 10–5 mol/L of SA that increased the activity of CA. Further, exogenous SA is also known to increase pigment concentration, activate Rubisco and PEP carboxylase (Singh and Usha 2003). Carbohydrates are translocated in bulk to the nodules for the use of rapidly metabolizing bacteria which is consistent with the increased carbohydrate content in nodules. However, the higher concentration of SA might have resulted in a permanent change at the level of membrane organization of the cells that proved to be injurious for plant's general metabolism and thereby reducing its overall growth and photosynthetic attributes, which is further supported by the findings of Uzunova and Popova (2000).

The application of lower concentration of SA elevated the activity of antioxidant enzymes viz. CAT, POX and SO which is in conformity with others (Hayat et al. 2008). This increase in the activity of antioxidant enzymes might be due to the regulatory role of SA at the level of transcription and/or translation. SA application is known to increase the endogenous proline content (Hayat et al. 2010) which is in conformity with the results of the present study where exogenous application of SA increased the *de novo* synthesis of proline thereby increasing its content in leaves. Proline besides acting as an excellent osmolyte is also known for stabilizing the complex II electron transport (Hamilton and Heckathorn 2001), membranes and 3-D structure of proteins (Holmström et al. 2000) and enzymes such as Rubisco and CA (Mäkelä et al. 2000) thereby increasing photosynthetic attributes. Thus the exogenous application of 10–5 mol/L of SA resulted in the increase in endogenous proline level, whereas higher concentration might have reversed the phenomenon.

The sequence that leads to the establishment of nitrogen fixing mature nodules begins with the infection thread formed by the joint action of bacteria and host, which is activated by phytohormones (Hopkins 1995). Exogenous application of SA favours the legume-*Rhizobium* symbiosis (Hayat et al. 2010) thereby leading to the enhanced establishment and development of nodules which is expressed in terms of increased number and dry mass of nodules. The healthy nodule development and increased leghemoglobin content will obviously lead to increased activity of nitrogenase, an oxygen labile enzyme which is protected by leghemoglobin. All these processes together bring about an efficient nitrogen fixation expressed in terms of increased nitrogen content.

The nitrogen fixing potential of each nodule is determined by three main factors (Cooper and Scherer 2012): (a) photosynthates availability, which was maintained by enhanced photosynthesis, low oxygen supply to the bacteroid, which at excessive level inhibits nitrogenase, was maintained by restricted O_2 supply by the mediation of increase in leghemoglobin export of fixed nitrogen in the form of ammonia. Nitrogen fixed in the form of ammonia diffuses across the peribacteroid membrane to the host cytosol by simple diffusion (Udvardi and Day 1990). Here two enzymatic systems, (a) GDH and (b) GS, GOGAT are operative to further metabolize it. GDH causes direct reductive amination of α-ketoglutarate, giving glutamate, whereas, GS catalyses the addition of NH_4^+ to glutamate forming corresponding amide, glutamine. This glutamine is converted back to glutamate by transfer of amide group to a molecule of α-ketoglutarate (Hopkins 1995). Since SA acts at transcriptional and/or translational level (Hayat et al. 2010), therefore might have accelerated the synthesis and thereby the activity of GDH, GS and GOGAT. Further, the activity of enzymes GS, GOGAT and GDH might also be increased by lower concentration of SA through the involvement of auxins, as auxins are known to enhance the activity of GS, GOGAT and GDH (Hayat et al. 2009). The application of 10–5 mol/L SA as a foliar spray increased the nitrogen fixation and assimilation and higher net photosynthetic rate which will probably increase the productivity of the legume crops.

The uptake of mineral elements by plant roots and their subsequent distribution within the plant have been the subject of studies for many decades (Marschner, 1995; Mengel et al., 2001; Karley and White, 2009; Miller et al., 2009; Miwa et al., 2009; Puig and Pen˜arrubia, 2009; White and Broadley, 2009). Mineral elements are acquired from the soil solution by plant roots. They traverse the root via apoplastic (extracellular) and/or symplastic (intracellular) pathways to the stele, where they are loaded into the xylem for transport to the shoot. All cytotoxic cations must be transported either through the apoplast or through the symplast in a chelated form. Several mineral elements are retained in the roots of some plant species. Examples are Ca, Mo, Na, Cd and Al (Marschner, 1995). The xylem delivers mineral elements to transpiring leaf tissues, where these are taken up from the apoplast by specific cell types. Recirculation of mineral elements within the plant, and the delivery of mineral elements to non transpiring or xylem-deficient tissues, occurs via the phloem. K, Na, Mg, Cd, N, P, S, Se and Cl are transported readily, but Fe, Zn, Cu, Mo and iodine (I) are less mobile, and Mn and Ca are essentially immobile in the phloem of most plant species (Marschner, 1995). The phloem mobility of B varies between species (Brown et al., 2002). Mineral elements that have low phloem mobilities accumulate in tissues with high transpiration rates and are present at low concentrations in fruits, seeds and tubers (Karley and White, 2009; White and Broadley, 2009; White et al., 2009). The ionome is defined as per Salt et al. (2008). It includes all mineral elements, whether essential or non-essential for life, in whatever chemical form these occur. Current interest is primarily focused on quantifying the ionome of different plant species, ecotypes and induced mutants, and on explaining differences in the ionome between and within plant species at the genetic level, although the ionome is also influenced profoundly by developmental and environmental factors (Watanabe et al., 2007; Baxter, 2009; Chen et al., 2009; White and Broadley, 2009; Broadley et al., 2010). It is self-evident that the uptake of mineral elements and their distribution within the plant determine the ionome of both the plant and its tissues. These processes are therefore fundamental not only to the mineral nutrition of the plant, but also for increasing the concentrations of mineral elements in edible tissues for human consumption. Plants appear to be composed of relatively few different cell types (Martin et al., 2001). Each of these cell types is thought to perform a distinct physiological function and, consequently, to have a unique ionome (Punshon et al., 2009; Karley and White, 2009). Recently, Conn and Gilliham (2010) review the phenomenon of cell-specific accumulation of mineral elements in plants. They describe the techniques used to determine both the tissue and subcellular distributions of mineral elements and present the tissue distributions of diverse elements including K, P, Ca, Na and Cd. They speculate on the physiological reasons for these distributions and the transport processes that are likely to generate them. They observe that the accumulation and tissue distributions of Ca, Na and Cd can be defined by the expression of key transport proteins and suggest how this phenomenon might be utilized to prevent the accumulation of toxic elements by plants. Similar observations are made by Miwa and Fujiwara (2010), who report that the misexpression of genes encoding B transporters alter the uptake and distribution of B in plants, and by Kobayashi et al. (2010), who report that misexpression of the iron deficiency-responsive cis-acting element binding factors 1 and 2 (IDEF1 and IDEF2) alters the expression of various genes encoding transport proteins that impact the uptake and distribution of Fe in plants. Indeed, the role of transport proteins in the uptake and distribution of mineral elements is highlighted by many articles in this Special Issue. These articles reveal that the expression of genes affecting transport processes influence plant adaptation to soils with extreme phytoavailabilities of mineral elements (Eticha et al., 2010; Fu¨hrs et al., 2010; Kobayashi et al., 2010; Miwa and Fujiwara, 2010), fertilizer use efficiencies in crops (Masclaux-Daubresse et al., 2010; White et al., 2010) and the biofortification of edible portions with essential mineral elements (Ding et al., 2010). Glass (1974) observed clear impact of phenolic acids on the uptake of regulation of various ions in the growth and development of plants. Phenolic acids altered membrane functioning and maintain the cellular pool ions specially potassium (K^+) which play a number of important roles in plant growth and developments.

2. Nutrient Balance and Plant Growth

A. *Nutrient Sensing and Signaling*

Plants are a diverse group of organisms distributed under many different environments in which they grow and to which they are adapted. Environmental change occurs due to location, seasonal variation, and daily, weekly, monthly, and yearly changes in climate. Plant diversity in genetic architecture, gene expression, and physiological adaptations is very large (Fu & Dooner 2002). Plants are anchored in one place for most of their life cycle and therefore must be able to adapt to a range of abiotic and biotic stresses. Variation in soil nutrient composition is found in agriculture systems and in natural environments and ranges from an extreme lack of nutrients in soils (e.g., Handreck 1997), to soils containing optimal amounts of fertilizer, and to soils with excess nutrients. To provide a framework for understanding plant adaptation to the range of nutrients found in soils, one may consider two different modes of plant growth and metabolism. One mode is when plants grow under nutrient-sufficient conditions where mineral resources are not limiting. Growth proceeds at optimal rates and it may be relatively easy for plants to acquire nutrients and water from soils where sunlight is abundant. When nutrients become limiting, growth is reduced and plants alter certain aspects of their acquisition, utilization, and morphology to maximize and acquire scarce resources. Plant species respond differently to nutrient limitations based on the environment in which they evolved (Grime 1977). Responses to nutrient limitations also depend on the nutrient, such as whether it is mobile or immobile in the soil (Hodge 2004). Crop plants have been bred over the past 30–40 years under high nutrient conditions and have high intrinsic growth rates and yield. Such characteristics may not be the most suitable for lower input environments. Plant response to the deprivation of essential mineral nutrients is fairly well described, but an understanding of how plants sense and signal changes (at the cellular or whole-plant level) in the low availability of nutrients. As fertilizer becomes more expensive and as farmers reduce fertilizer usage because of the negative environmental impacts, it will be important to gain a better understanding of how crop plants can be designed to grow more efficiently in environments with lower nutrient inputs (Good et al. 2004). In addition, access to fertilizers in developing countries is limited (Sanchez 2002), so it could be helpful to create higher-yielding crop plants that are adapted to a lower input environment.

To adopt and grow in nutrient deprived environment plants must sense changes in external and internal mineral-nutrient-concentrations and adjust growth through phytohormonal cross-talk. Because plant adaptation to macronutrient deprived environment is based on nutrient sensing and integration of tissue nutrient concentrations. Sensor molecule could be affected or activated by indirect changes in plant metabolism that result from stress such as metabolic changes, ionic imbalance, accumulation of ROS, release of ATP or reduced energy levels (Mittler and Blumwald, 2010).

It was recently shown that energy depletion in *Arbidopsis* during abiotic stress is directly linked to the activation of abiotic stress responses in $SnRK_1$ [Upstream regulator of stress response in plants]. This upstream regulator triggers extensive transcriptional changes involving over 600 genes and contributes to the restoration of cellular homeostasis and survival. The large number acclimation responses and signals controlled by $SnRK_1$ demonstrate that this protein functions as a key regulator of abiotic stress responses in plants. Another putative stress receptor that appears to function at high level along the osmotic stress response signaling pathway is $ATHK_1$. Recently a protein is isolated to function as an upstream regulator of salinity responses in *Medicago truculata* is a novel leucine-rich repeat receptor kinase(Srlk) (Deloranzo *et al.* 2009).

Nitrogen and carbon metabolism are linked and therefore cross talk between the signal transduction pathways that regulate nitrogen assimilation and carbon metabolism is expected. Systems analysis provides new insights that should help to further elucidate these complex networks (Palenchar et al. 2004, Stitt et al. 2002). To date, much work on nitrogen sensing has focused on the metabolic and morphological responses to the addition of nitrate, in particular the induction of nitrogen metabolism (Scheible et al. 2004, Stitt et al. 2002, Wang et al. 2000, Wang et al. 2003, Wang et al. 2001) and long-distance signaling and transporter responses (Forde 2002). The root responses to nitrate are well documented, and even though the transporterNRT2.1 was reported to be involved in nitrate sensing or transduction (Little et al. 2005) in roots, a more recent report suggests an alternative role for NRT2.1 (Remans et al. 2006). Plant responses and signal transduction pathways for nitrogen limitation are poorly understood because experiments that focus on the switch from nitrogen-sufficient to -deficient conditions have received less attention. However, the importance of nitrogen limitation cannot be overstated as nitrogen-use efficiency is a critical future goal for agriculture because of the need for enhanced stewardship of the environment and for more efficient use of an increasingly expensive input (Gallais & Hirel 2004). PII is potentially a key protein involved in the sensing of internal nitrogen supply. PII is a homolog of the well-characterized *Escherichia coli* protein involved in the nitrogen regulatory system (Moorhead & Smith 2003, Hsieh et al. 1998). The protein PII in *E. coli* is encoded by the *GlnB* gene, which interacts with other proteins to regulate glutamine synthase (GS). In plants, much of carbon and nitrogen metabolism occurs in the chloroplast where PII is localized (Moorhead & Smith 2003). The plant PII homolog interacts with N-acetyl glutamate kinase, which is a key enzyme specifically involved in arginine biosynthesis (Chen et al. 2006, Takei et al. 2002) and in nitrogen metabolism in general. Nitrate may be a key molecule that is sensed by plants (as with phosphate) and is involved in controlling the ratio of roots to shoots (Scheible et al. 1997). A highly significant correlation was found between leaf nitrate and shoot-root ratios

(Scheible et al. 1997). Another important factor in root-to-shoot allocation that is influenced by nitrogen is the hormone cytokinin (Ferrario-Mery et al. 2006). Cytokinin levels increase when plants are supplied with nitrogen and decrease when deprived. Cytokinins play a potentially important role in root-to-shoot communication as they are synthesized in roots and are well correlated with observed changes in biomass allocation between roots and shoots (Sakakibara 2006). Possible response regulators downstream of a GS: glutamine synthase cytokinin receptors have been identified, supporting a role for cytokinin in nitrogen responses (Sakakibara et al. 1998, Taniguchi et al. 1998). Under nutrient-deprived conditions, phosphorylation also plays a key role in fine-tuning plant responses. In wheat, a gene encoding an *SNF1* kinase is up-regulated by nitrogen, phosphorus, and sulfur deficiency (Sano & Youssefian 1994). This kinase is also induced by cytokinin, even though cytokinin levels would be expected to decline under nitrogen deficiency. Another phosphorylation event and potential signaling cascade regulated by nitrogen deprivation is the phosphorylation and conversion of the nitrate transporter CHL1 to a high-affinity mode (78 Liu & Tsay 2003). The kinase that phosphorylates CHL1 has not been identified but will be an important component for understanding the signal transduction cascade in response to nitrogen deprivation. Phosphylation and dephosphorylation are also known to activate and inactivate nitrate reductase (Kaiser & Huber 2001). More work is needed to identify the networks of phosphorylated proteins and their downstream targets involved in the nitrogen signal transduction cascade. In addition to sensing nitrate, plant cells may also sense carbon status, which leads to the regulation of key nitrate transporters *NRT2.1* and *NRT1* (Filleur et al. 2005). A key metabolite involved in nitrogen metabolism is glutamate. The presence of multiple glutamate receptors in plant genomes (Lacombe et al. 2001) may indicate the potential importance of this metabolite in initiating signal transduction cascades. However, the role of glutamate receptors or glutamate is only beginning to be elucidated. Antisense lines of *AtGLR1.1* displayed conditional phenotypes in response to changes in carbon to nitrogen ratios. A model was presented that suggests that this receptor regulates abscisic acid (ABA) biosynthesis during seed germination, which is stimulated by nitrate and inhibited by sucrose (Kang & Turano 2003). Glutamate also leads to changes in cation fluxes (Demidchik et al. 2004) as well as changes in root growth (Lejay et al. 1999, Walch-Liu et al. 2006). At this time these putative receptors cannot be placed in signal transduction cascades. Glutamine is another key metabolite linked to nitrogen metabolism that plays roles in metabolic regulation and possibly signal transduction (Sugiharto et al. 1992). Some indications for possible signal transduction networks can be found in studies on global transcriptional changes after two days of nitrogen deprivation and MADS-box transcription factors after 2.5 days of nitrogen deprivation (Gan et al. 2005, Scheible et al. 2004). Global transcriptional changes to nitrogen deficiency were mainly described in terms of changes that occurred after reintroduction of nitrate (Scheible et al. 2004), and therefore discussion of this rich data set on the transcriptional changes in response to nitrogen deprivation is beyond the scope of this review. However, extensive changes were found after two days of nitrogen deprivation, including a large number of transcription factors and kinases (see the supplemental table in Reference Scheible et al. 2004). A previously identified MADS-box transcription factor, *ANR1*, which was identified as being induced by nitrate, is actually up-regulated by nitrogen deprivation and not rapidly repressed by nitrate (after 3 h of nitrate) (Gan et al. 2005). Seven other MADSbox transcription factors are also up-regulated by nitrogen deprivation, but to a lesser extent (Gan et al. 2005). A Myb transcription factor that is a member of the *PHR* family of transcription factors in plants (Rubio et al. 2001) was found to be upregulated by nitrogen deprivation (Todd et al. 2004). These Myb factors likely play important roles in the signaling of nitrogen deprivation. Other Mybs in the R2R3 family have also been shown to be responsive to nitrogen deprivation (Miyake et al. 2003), but their targets or physiological roles are unknown. Gene expression data provide potential components of signal transduction pathways, but additional work will be required to link these components into networks.

Salicylic acid (SA) is a phenolic compound which, despite its broad distribution in plants, has basal levels differing widely among species, with up to 100-fold differences having been recorded (Raskin et al., 1990). This disparity can be observed within members of the same family. For example, in the Solanaceae, whereas tobacco (*Nicotiana tabacum*) contains low basal levels of SA [<100 ng g_1 fresh weight (FW)] in leaves (Yalpani et al., 1991; Malamy et al., 1992), potato (Solanum tuberosum) might contain up to 10 lg of total SA g_1 FW (Coquoz et al., 1998; Navarre and Mayo, 2004). In the model plant Arabidopsis thaliana, basal levels of total SA range from 0.250 lg to 1 lg g_1 FW (Nawrath and Me´traux, 1999; Wildermuth et al., 2001; Brodersen et al., 2005). SA is synthesized through two distinct and compartmentalized pathways that employ different precursors: the phenylpropanoid route in the cytoplasm initiates from phenylalanine, and the isochorismate pathway takes place in the chloroplast. Most of the SA synthesized in plants is glucosylated and/or methylated. Glucose conjugation at the hydroxyl group of SA results in formation of the SA glucoside [SA 2-O-b-D-glucoside] as a major conjugate, whereas glucose conjugation at the SA carboxyl group produces the SA glucose ester in minor amounts. These conjugation reactions are catalysed by cytosolic SA glucosyltransferases that are induced by SA application or pathogen attack in tobacco and Arabidopsis plants (Lee and Raskin, 1999; Song, 2006). SAG is actively transported from the cytosol into the vacuole of soybean and tobacco cells, where it may function as an inactive storage form that can release free SA (Dean and Mills, 2004; Dean et al., 2005). Interestingly, SA is also converted to methyl salicylate (MeSA) by an SA carboxyl methyltransferase, and this volatile derivate is an important long-distance signal in tobacco and Arabidopsis systemic acquired resistance (Shulaev et al., 1997; Chen et al., 2003; Park et al., 2007; Vlot et al., 2008). MeSA can be further glucosylated to produce MeSA 2-O-b-D-glucose, but this SA-conjugated form is not stored in the vacuole (Dean et al., 2005). The reader is referred to excellent reviews dealing with the enzymes and regulation of these biosynthetic routes (Klessig and Malamy, 1994; Lee et al., 1995; Shah, 2003; Chen et al., 2009; Vlot et al., 2009).

B. *Nutrient Transportation/Mobilization*

To attain optimal growth and development, plants tend to maintain constant level of essential mineral nutrients, despite their limited availability in most soils. These limitations are due to low nutrient concentration or accessibility. Thus mainly nutrients have low solubility or present an irregular, patchy distribution in soils. To cope with reduced nutrient availability, plants trigger physiological and developmental responses aimed to increased nutrient acquisition (Rubio *et al.* 2009). These responses include induced expression of specific high affinity transporters and accumulation and/or secretion of enzymes and compounds that facilitate nutrient remobilization from inorganic and organic sources.

Nitrogen (N) is the basic macronutrient required for growth and development of plants. Plants take up N primarily as nitrate and ammonium, with nitrate being predominant form in most agricultural soils. The function of several structural genes involved in N uptake and assimilation have been studied in the past. In *Arabidopsis* there are three families of nitrate transporters NRT1 NRT2 and CLC with number of genes (Kant et al. 2011). The NRT1 family members are low affinity nitrate transporter while NRT2 are high affinity nitrate transporters. Nitrate after entering the plant cell, is reduced to nitrate by nitrate reductase and further to ammonium by nitrite reductase which is assimilated into amino acids via GS/GOGAT pathway (Crawford and Forde 2002).

The leaves are a sink for N during the vegetative stage and, afterwards, this N is remobilized for use in the developing seeds. The majority of this remobilization occurs during senescence whereto N is transported mainly via amino acids. Up to 80% of grain N contents are derived from leaves in rice and wheat (Kichey et al., 2007; Tabuchi et al., 2007). Plants have developed efficient methods and mechanisms that release tied-up N entities from source tissues via protease activities during leaf senescence. The protein degradation occurs mainly by three pathways: the chloroplast degradation pathway; the vacuolar and autophagic pathway; and the ubiquitin-26S proteasome pathway (Liu et al., 2008). Approximately 80% of total leaf N is located in the chloroplasts mainly in the form of proteins and this is an important N pool for remobilization (Adam et al., 2001). Among chloroplastic proteins, Rubisco (;50% of total cellular proteins in C3 and ;20% in C4 plants) seems to serve as the major protein subjected to proteolysis and responsible for most N remobilized during leaf senescence for grain-filling (Mae et al., 1993).

In higher plants, NO_3^- assimilation is dependent on the supply of carbon skeletons, which indicates a close interaction between carbon and N metabolism. NO_3^- assimilation proceeds at a low rate in plants with low carbohydrate levels (Stitt et al. 2002). Certain sugars increase the NO_3^- assimilation rate and amino acid synthesis (Morcuende et al. 1998). The results of studies on mustard (*B. juncea* Czern & Coss cv. Varuna) and wheat (*Triticum aestivum* L.) have demonstrated a direct relationship between photosynthetic CO_2 assimilation and NO_3^- assimilating enzymes in response to SA (Fariduddin et al. 2003; Singh & Usha 2003). In these studies, plants were treated by foliar application of SA; in the present study, we instead tested pre-soaked seeds with SA in the absence and presence of NO_3^-. The

NO_3^- assimilation rate increased in cotyledons in response to 50 mM SA, together with an increase in the amount of soluble sugars and starch, though the increase in starch content was low if compared with that of total N in cucumber. The effect of exogenous SA on the physiological characteristics of plants may depend on the SA concentrations as well as on the nutritional conditions of the plants. The present study points to a positive correlation between chlorophyll content and total N in cucumber cotyledons. Moreover, the effects of SA were more significant in the absence than in the presence of NO_3^-. The increase in N and chlorophyll content at lower SA concentrations indicates that SA plays a regulatory role during the biosynthesis of active photosynthetic pigments. Although the direct effect of SA on chlorophyll biosynthesis in plants is not clear, a-ALA-mediated enhancement in chlorophyll biosynthesis by benzyladenine (synthetic SA) has been reported (Ananiev et al. 2004). The reduction we observed in the level of total N and chlorophyll content at 500 mM SA may be due to the breakdown/degradation of chlorophyll or due to the inhibition of foliar proteins, which are required for the production of photosynthetic pigments.

C. *Nitrogen Use Efficiency in Plants*

NUE has been defined in various ways (Good et al., 2004), but the simplest is the yield (grain, fruit or forage) per unit of N available in the soil. There are two general stages for N use in the plant life cycle. First, during biomass formation there is the amount of N uptake, storage, and assimilation into amino acids and other important nitrogenous compounds. The second stage is the proportion of N that is partitioned to the seed, resulting in final yield. During the vegetative stage, young developing leaves and roots behave as sinks for inorganic N uptake and synthesis and the storage of amino acids via the nitrate assimilation pathway. These amino acids are further utilized in the synthesis of proteins and enzymes involved in different biochemical pathways and the photosynthetic machinery governing plant growth, architecture, and development. Later on, during the reproductive stage the increased supply of nitrogenous compounds is necessary for optimum flowering and grain filling. At this stage, both N assimilation and remobilization become critical and the leaves and shoot act as the source providing amino acids to the reproductive and storage organs. Therefore, understanding the mechanisms for N uptake, assimilation, and remobilization during the plant life cycle is important for increasing NUE. Nitrate is not only the predominant source of N supply to plants, but also acts as an important signal for several developmental processes. This regulation includes a rapid change in expression pattern of genes involved in carbon (C) and N metabolism and other metabolic pathways. Further, its concentration affects root development, root architecture, and the root-to-shoot ratio. Interestingly, not much is known about the molecular mechanisms and regulatory genes that govern these nitrate responses, although some transcription factors and kinases have been linked to these processes.

Development of genetic varieties with improved nitrogen use efficiency (NUE) is essential for sustainable agriculture. One important approach is to develop and understanding of the plant responses to N limitation and its regulation through metabolic homeostasis. There are several genetic and environmental factors which might limit NUE such as root architecture, C/N ratio in the soil, accumulation of phenylpropanoids (Kant *et al.* 2011; Singh and Chaturvedi 2012). Acomplementary approach would be to unravel the molecular mechanisms for identifying unique and regulatory genes and promoter elements in plant species that are naturally N limitation. Therefore, working with a model plant that grows better under N stress conditions would facilitate investigation of the molecular mechanisms regulating adaptation to low N that may confer an improved NUE.

Thallungiella halophile has emerged as a new model plant for the molecular elucidation of abiotic stress tolerance. *Thallungiella* stress similar morphology and sequence identify with *Arabidopsis* thus allowing for the utilization of *Arabidopsis* genetic information to investigate *Thalungiella* responses to stress. Also, it has all the attributes of a model plant system including a short life cycle, a relatively small genome copious seed production, and ease of transformation (Kant *et al.* 2011). *Thallungiella* have various nitrate transporter genes and N assimilatory enzymes compared with *Arabidopsis* under both N-sufficient and N-deficient conditions.

Plants have evolved adaptive responses that allow them to grow in soils even in the absence of certain nutrients. These responses entail complex metabolic changes generated by nutrient deficiency. In this study, we demonstrate that SA induces different responses depending on the concentrations applied. A concentration of 50 mM SA seems to provide an optimal balance between the nutrient level and the growth and development against stress. By contrast, higher doses of SA induced inhibitory responses in cucumber plants. Indeed, the germination rate was significantly higher at 50 mM SA than that observed in response to higher doses of SA both in the absence and presence of exogenous NO_3^-. Higher doses of SA may inhibit NO_3^- uptake and cause retardation in growth and development. Glass (1974) observed concentration-based inhibition potency of PAs on ion uptake in barley (*Hordeum vulgare* L. cv. Karlsberg). Rajjou et al. (2006) have reported similar observations on seed germination and seedling establishment of Arabidopsis thaliana. SA might induce the mobilization of internal tissue NO_3^- and chlorophyll biosynthesis to increase the functional state of the photosynthetic machinery in plants (Shi et al. 2006), or it may result in the accumulation of à amino levulinic acid (à -ALA) in cotyledons. Ananiev et al. (2004) reported an increase in chlorophyll biosynthesis in excised cotyledons of *Cucurbita pepo* L. (zucchini), cv. Cocozelle in response to growth regulators. This inductive process may be due to the interaction of PAs with light (McClure 1997; Hemm et al. 2004), which might induce higher rates of carbohydrate synthesis through photosynthetic activity; this finding may be ascribable to changes in membrane organization at a higher SA dose or to chelation of some important elements in the membrane of cell organelles (Uzunova & Popova 2000). It is not clear why N content increased when exogenous $NO_3^-$was not supplied. Internal tissues NO_3^- probably induce NR activity' at lower concentrations of SA and/or SA induced modulation of nitrogen use efficiency (NUE) in cucumber cotyledons (Singh & Singh 2008). The increase in NO_3^- assimilation may be related to the physiological concentration (50 mM) of SA under poor exogenous NO_3^- supply conditions. The imbalance between demand and N supply in crops can result in a sub-optimal yield and/or environmental damage due to the addition of an excessive amount of fertilizers. NO_3^- uptake and assimilation by roots are known to be influenced by N supply, which suggests that the N status of plants might be influenced, and feedback to regulate these processes with interactions of phytohormones (Rubio et al. 2009). Limited N availability reduces growth and plant productivity and induces secondary metabolism (Chisaki & Horiguchi 1997; Lattanzio et al. 2009). Our HPLC data support the hypothesis that SA favors growth and development by increasing NUE in cucumber. In the absence of NO_3^-, SA accumulation plays a protective role against nutritional disorders. One report (Vicente & Plasencia 2011) confirms our finding that the exogenous application of 50 mM SA enhances growth and development to a greater extent than high SA doses (500 mM). One possible explanation for the concentrationbased effect of SA on NR activity is that enzyme activity is induced and/or protected from degradation. Our results suggest that 10–50 mM SA induces NR synthesis by mobilization of intracellular NO_3^-, thereby providing protection against *in vivo* NR degradation in the absence of NO_3^- (Singh et al. 1997). Fariduddin et al. (2003) reported increased NR activity due to reduced concentrations of SA, whereas high concentrations inhibited NR activity, in *Brassica juncea* Czern and Coss cv. Varuna.

D. *Nutrient Scarcity & Oxidative Enzymes*

Understanding plant response to nitrogen limitation for the improvement of crop nitrogen use efficiency (Journal of Experimental Botany, 2011)

Plants in natural field conditions face changing environmental conditions where N concentrations vary and frequently are limiting for growth due to many factors including surface run-off/soil erosion, rainwater leaching, gaseous losses by volatilization, and microbial consumption. Therefore, adaptation to limiting N conditions is an important survival strategy for plants successfully to complete their life cycle. For crops grown in developed countries, the use of large amounts of N fertilizer for many crops helps prevent fluctuating levels of N from impacting yield and, as a consequence, much is wasted to the environment. In developing countries, many farmers cannot afford to use much N fertilizer. Therefore, in either case developing crops that have improved genetics for yielding well under limiting N conditions would be very advantageous. In our laboratory, research has been focused on understanding the response of plants to limiting N and to try to use this knowledge to improve NUE either by increasing yield under existing levels of N supply or by maintaining yield by decreasing N levels. Most of the N transporters, assimilatory, and regulatory genes in plants have been first identified in the model plant Arabidopsis. However, Arabidopsis is not adaptive to low-N conditions, since its growth is retarded sharply under changing N availability (Martin et al., 2002; Peng et al., 2007a, b; Kant et al., 2008). A complementary approach would be to unravel the molecular mechanisms for identifying unique N regulatory genes and promoter elements in plant species that are naturally adapted to N limitation. However, it has so far not been easy to work with plants that show better growth under N-limiting conditions because of their genetic complexity and the lack of molecular and genetic tools available in these plants. For example, modern maize hybrids can grow better and produce higher yield than older lines under limited-N conditions and variations among rice, wheat, barley, and sorghum cultivars for NUE are known (Muchow, 1998; Anbessa et al., 2009; Namai et al., 2009). However, little is known about the molecular mechanisms leading to higher NUE in these newer hybrids and cultivars. Therefore, working with a model plant that grows better under N stress conditions would facilitate investigation of the molecular mechanisms regulating adaptation to low N that may confer an improved NUE.

Exposure of plants to unfavorable environmental conditions such as drought, salinity, ozone depletion and nutrient unavailability can increase the production of reactive oxygen species. During salinity-induced oxidative stress, two primary processes are involved in the formation of ROS during hotosynthesis. One is the direct photoreduction of O_2 to the superoxide radicle by reduced electron transport components associated with PS I (Mehler reaction), and other are the reactions linked to the photorespiratory cycle, including Rubisco in the chloroplast and glycolate-oxidase and CAT peroxidase reactions in the peroxisome. These processes initiate chain reactions that produce more harmful oxygen radicals (Hsu and Kao 2003). These cytotoxic reactive oxygen species (ROS) can destroy normal metabolism through oxidative damage of lipids, proteins, and nucleic acids when they are produced in excess (McCord 2000). Plants have evolved efficient systems for ROS removal, which include specific ROS-scavenging enzymes like superoxide dismutase (SOD; EC 1.15.1.1), ascorbate peroxidase (APX; EC 1.11.1.11), catalase (CAT; E.C 1.11.1.6), glutathione reductase (GR; EC 1.6.4.2) and glutathione peroxidase (GPX; EC 1.11.1.9). and small non-enzymatic molecules that act as ROS scavenger such as ascorbate, glutathione, tocopherol, flavonoids, anthocyanines, polyphenolic compounds and carotenoids. ROS are regularly synthesized in the chloroplast (Asada 1992) and mitochondria (Rich and Bonner 1978), though some quantity is also reported to be produced in microbodies (Lindquist *et al.* 1991). SOD is a metalloenzyme that catalyze the disproportionation of toxic superoxide radicles to hydrogen peroxide (H_2O_2) and O_2 and thus play a very crucial role in the maintenance of low concentration of toxic radical. SODs occurs in three different molecular forms which contain copper and zinc (CuZn) iron (Fe) or manganese (Mn) as prosthetic metals. CuZn-SODs are the major enzymes in plants and are known to be localized in chloroplast (Asada 1992), cytosol (Bowler *et al.* 1994) and peroxisomes of oil seeds (Corpas *et al.* 1998). Mn SODs occur mainly in mitochondria and are also found in peroxisomes and in chloroplasts while Fe SODs have frequently been shown to be present in prokaryotes but only rarely in the chloroplasts of higher plants. Peroxidases (EC. 1.11.1.7) function as effective quenches of reactive intermediary forms of oxygen and peroxy radicals produced under stress (Zhang and Kirkham 1994). Peroxidases form large families with substrate specificity, its function are ubiquitous in plants, and catalyze the oxidation of cellular components by either H_2O_2 or hydroperoxides. Ascorbate Peroxidase that uses ascorbate as its reducing substrate and is the major H_2O_2 scavenging enzyme in plants under normal and stress conditions and occur as several isozymes localized in different cellular compartments like cytosol microbodies, chloroplasts, etc. (Madhusudhan *et al.* 2003). Enzyme which requires guaiacol (2-methoxyphenol) is known as guaiacol peroxidase (GPox EC 1.11.1.7). There is a significant reactivity differences between ascorbate specific and guaiacol specific Peroxidase. They also differ in their physiological roles. Reaction catalyzed by APox prevents accumulation of excess H_2O_2 and this provides for cell defense against oxidative stress (Yoshimura *et al.* 2000). In contrast guaiacol peroxidases also exist in several isoenzymic forms and are located in cytosol, vacuole, cell wall, extra cellular space, oxidize a wide range of organic substrates and are involved in processes like lignification of cell wall as well as in wound healing (Gazaryan *et al.* 1996). The role of GR in H_2O_2 scavenging in plant cells has been well established in Halliwell-Asada enzyme pathway (Bowler *et al.* 1992). Elevated levels of GR activity perhaps increase the ratio of $NADP^+/NADPH$ thereby ensuring availability to $NADP^+$ to accept electron from photosynthetic electron transport chain, minimizing the formation of GSH/GSSG helps in regeneration of ascorbate which is necessary for the activation of several chloroplastic CO_2 fixing enzymes and accumulation of GSH that ultimately confers tolerance to stress in plants (Baisak *et al.* 1994). The harmonized activities of the multiple forms of these enzymes in different subcellular compartments achieve a balance between the rate of formation and removal of ROS, and sustain H_2O_2 at the required levels for cell signaling. It is now widely accepted that the degree of oxidative cellular damage in plants exposed to abiotic stresses is controlled by the capacity of the antioxidative systems (Acar *et al.* 2001; Türkan *et al.* 2005). A correlation between antioxidant capacity and salinity tolerance has been reported in several plant species such as rice (*Oryza sativa* L.)(Demiral and Türkan 2004, 2005), wheat (*Triticum aestivum* L.) (Meneguzzo *et al.* 1999), pea (*Pisum sativum* L.) (Hernández *et al.* 1995, 2000), sesame (*Dalbergia sissoo* L.) (Koca *et al.* 2007), Indian mustard (*Brassica juncea* L.) (Yusuf *et al.* 2008), corn (*Zea mays* L.) (Szalai and Janda 2009). Furthermore, transgenic plants overexpressing ROS-scavenging enzymes, such as SOD (Breusegem *et al.* 1999; Alscher *et al.* 2002), APX (Wang *et al.* 1999), GR (Foyer *et al.* 1995) and GPX (Roxas *et al.* 1997, 2000) showed enhanced tolerance to osmotic, temperature, photoinhibition and oxidative stresses. Transgenic tobacco plants overexpressing APX in their chloroplasts showed enhanced tolerance to salt, PEG and water stresses (Badawi *et al.* 2004) and those expressing glutathione-*S*-transferase constitutively showed protection against both cold and salt stresses (Roxas *et al.* 1997). Mohamed *et al.* (2003) has recently shown that tomato plants that are expressing bacterial catalase gene in their chloroplasts were protected against the photo-oxidative stress caused by paraquat treatment, drought stress and chilling stress.

3. SA & Nutrient Stress Management Under Changing Environment

Investigations show the role of SA in modulating plant responses to a wide range of osmotic/ and or oxidative stresses which revert the negative impact of stresses (Singh *et al.* 2011). Concentration based regulation of acclimation to abiotic-stresses by exogenous application of SA in plants may become a specific tool to understand the physiology, molecular biology and biotechnology of plants. Plants respond to stress by the synthesis of signaling molecules. These activate a range of signal transduction pathways. Several such signaling molecules have been identified in plants as calcium, jasmonic acid, ethylene and SA. The role of SA as a defense signal has been well established in plants (Klessig and Malamy 1994). SA has qualified as a plant hormone due to its physiological and biological roles in plants (Raskin 1992). SA has been suggested as signal transducer or messenger under stress conditions (Klessig and Malamy 1994). Numerous studies have indicated that exogenous application of salicylic acid to the stressed plants can potentially alleviate the toxic effects, generated by stress factors. For example an enhanced tolerance against salinity stress was observed in tomato plants raised from the seeds soaked in salicylic acid and was presumed to be due to the enhanced activation of some enzymes viz. aldose reductase and ascorbate peroxidase and to the accumulation of certain osmolytes such as proline (Tari *et al.* 2002, 2004; Szepesi *et al.* 2005). The maintenance of growth and development in saline environments is associated with osmotic adjustment, and with the synthesis of osmoprotectants/ compatible osmolytes or in some halophytes, with the sequestration and accumulation of Na^+ in the vacuole (Ramani *et al.* 2006). Accumulation of large amounts of osmolytes (proline) is an adaptive response in plants exposed to stressful environments (Rai 2002). Wheat seedlings accumulated large amounts of Pro under salinity stress, which was further, increased when salicylic acid was applied exogenously, thereby alleviating the deleterious effects of salinity (Shakirova *et al.* 2003). The exogenous application of salicylic acid prevented the lowering of IAA and cytokinin levels in wheat plants resulting in the betterment of cell division in root apical meristem, thereby increasing growth and productivity of plants (Shakirova *et al.* 2003). These authors also reported that the pre-treatment with SA resulted in the accumulation of ABA, which might have contributed to the pre-adaptation of seedlings to salinity stress as ABA induces the synthesis of a wide range of anti stress proteins, thereby providing protection to the plants. Further, the treatment also lowered the level of active oxygen species and therefore the activities of SOD and POX were also lowered in the roots of young wheat seedlings (Shakirova *et al.* 2003). These findings indicate that the activities of these antioxidant enzymes are directly or indirectly regulated by salicylic acid, thereby providing protection against abiotic stress (Sakhabutdinova *et al.* 2004). The possible existence of positive amplification loops involving NADPH oxidases in ROS signaling has been suggested in several studies (Dat *et al.* 2003; Rizhsky *et al.* 2004). These loops might be activated by low levels of ROS and result in enhanced production and amplification of the ROS signals in specific cellular locations. The accumulation of ROS in cells might activate the ROS scavenging pathways and result in the suppression of ROS in specific cellular locations or the entire cell. Overmyer *et al.* (2003) proposed that SA might be involved in this positive amplification loop and amplifies signals leading to defense responses.

Plants have evolved adptive responses that allow them to grow under adverse environmental conditions. Both biotic and abiotic stresses reduced the nutrient absorption ability and its proper transportation/mobilization in plants. These responses implicate complex physiological and developmental changes that increase nutrient scavenging and remobilization capacity and protect the plants from the stress generated by nutrient deficiency. The existence of interaction between hormone and nutrient starvation signaling is very important to understand the nutrient stressed conditions in plants (Rubio *et al.* 2009). Recently, large amount of data have been generalized on the effects of hormone and nutrient starvation treatments in the transcriptome of *Arabidopsis*. Studies on the effect of nutrient starvation treatments on expression of hormone signaling and metabolism genes show up or down regulation in plants. A significant correlation for either up or down regulated genes for many of the hormones and nutrient starvation are examined (Table 1).

Table 1. Known role of hormones in the control of nutrient homeostasis.

Hormone	Role in nutrient signaling plant adaptation
SA	Up-regulation of nitrate reductase activity plant under scarcity of NO_3^- nutrition.
JA	Putative modulation of the expression of suffer metabolism genes possible role in recycling under potassium starvation stress.
ABA	Lateral root inhibition by nitrate resupply.
Auxin	Bidirectional antagonistic effect with sulfur deprivation signaling.
CK	Long distance (systemic) signal in nitrate resupply.
Ethylene	Root hair proliferation and elongation in response to low iron supply.

The role of SA in plant growth has been little studied compared with other plant hormones. Most reviews on this topic do not include SA, or its role is barely described (Santner and Estelle, 2009; Santner et al., 2009; Wolters and Ju¨rgens, 2009).The effect of exogenous SA on growth depends on the plant species, developmental stage, and the SA concentrations tested and or accumulated under adverse environmental conditions. Growth-stimulating effects of SA have been reported in soybean (Gutie´rrez-Coronado et al., 1998), wheat (Shakirova et al.,

2003), maize (Gunes et al., 2007), and chamomile (Kova´cik et al., 2009). In soybean plants treated with 10 nM, 100 lM, and up to 10 mM SA, shoot and root growth increase; 20% and 45%, respectively, 7 d after application. Wheat seedlings treated with 50 mM SA develop larger ears, and enhanced cell division is observed within the apical meristem of seedling roots (Shakirova et al., 2003). Likewise, 50 lM SA stimulates the growth of leaf rosettes and roots of chamomile plants by 32% and 65%, respectively, but higher concentrations (250 lM) have the opposite effect (Kova´cik et al., 2009). It has been suggested that the growth-promoting effects of SA could be related to changes in the hormonal status (Shakirova et al., 2003; Abreu and Munne´-Bosch, 2009) or by improvement of photosynthesis, transpiration, and stomatal conductance (Stevens et al., 2006). In *A. thaliana*, exogenous SA (100 lM and 1 mM) has a negative effect on trichome development because its application reduces trichome density and number (Traw and Bergelson, 2003). Although the biochemical events involved in the regulation of cell division and growth by SA are still unknown, these results correlate well with the anti proliferative properties in mammalian tumor cell lines of the acetylated derivative (Ru¨schoff et al., 1998; Dihlmann et al., 2001). Relationship between the SA signaling pathway and Arabidopsis growth rate. More direct evidence supporting the key role of endogenous SA in the regulation of plant cell growth comes from the characterization of *Arabidopsis* mutant or transgenic plants affected in the SA signaling pathway. *Arabidopsis* plants that overexpress the SA-inducible DOF (DNA binding with one finger) transcription factor OBP3 show a decreased growth rate in both roots and aerial parts of the plants, which in the most severe cases led to death (Kang and Singh, 2000). This dwarf phenotype is also observed in *Arabidopsis* mutants that have constitutively high levels of SA, such as cpr5 (constitutive expressor of PR5; Bowling et al., 1997), acd6-1 (accelerated cell death; Rate et al., 1999), and agd2 (aberrant growth and death; Rate and Greenberg, 2001). In contrast, the SAdepleted Arabidopsis NahG transgenic plants have a higher growth rate (Abreu and Munne´-Bosch, 2009; Du et al., 2009) that is reflected by a 1.7-fold increase in leaf biomass when compared with wild-type plants (Abreu and Munne´-Bosch, 2009). The effects of SA depletion on plant growth are more evident at low temperature; Arabidopis NahG transgenic plants grow faster at 4 _C than wild-type plants and show a similar growth phenotype to the amp1 mutant that has increased CK levels (Xia et al., 2009). The elevated CK levels or the decreased SA levels improve plant growth at low temperatures through different mechanisms. The higher growth rate of amp1 in the cold is associated with a continuous cell division rather than enhanced cell expansion, whereas the increased growth of NahG plants at 4 _C results from enhanced cell expansion rather than continuous cell division (Scott et al., 2004; Xia et al., 2009). Moreover, the enlarged cell size of NahG plants is associated with an extensive endoreduplication. NahG plants have approximately one additional endocycle compared with wild-type plants, resulting in DNA values as high as 32C. It has been suggested that SA negatively regulates expression of cyclin D3 (CYCD3; which drives the G1/S phase transition) because an increased expression is found in NahG plants grown at 4 _C (Xia et al., 2009). These results suggest an unexplored cross-talk between SA, CK, and BR signalling pathways since the latter two are positive regulators of CYCD3 expression (Riou-Khamlichi et al., 1999; Hu et al., 2000). Although most of the evidence suggests that SA is a negative regulator of cell division, its role is much more complex. Depletion of SA levels through NahG transgene expression reverts the acd6-1 phenotype, but lead to the appearance of abnormal tumour-like growths in the agd2 mutant background. The same effect is also observed in acd6 NahG plants treated with the SA analogue benzothiadiazole S-methylester (BTH; Rate et al., 1999). In acd6-1 mutants, SA stimulates endo reduplication and cell enlargement, while in the agd2 background SA suppresses both processes. This discrepancy could possibly be the result of SA interaction with multiple receptors or signaling pathways that control cell growth and development.

4. Conclusion & Future Perspective

In order to use gene knowledge to improve NUE, it is important to develop a more complete understanding of how different plants respond to growth under different N conditions. We have only begun to understand the regulatory mechanisms that underpin these responses. Considerable work still needs to be done using model systems like *Arabidopsis* and *Thellungiella* to improve this understanding. However, to improve this trait in important crops it is necessary to use this knowledge to improve breeding selection or through the development of transgenic lines. Many important crop traits like NUE are traditionally considered to be complex multi-gene traits, with no single gene contributing more than a small percentage to the phenotype. Therefore, it was unclear whether a single transgene could have a significant enough effect on phenotype to make commercialization worthwhile. However, there are several examples where a single transgene has had a significant phenotypic effect and although none have yet been commercialized it gives hope that this approach will be worthwhile. Certainly this approach will require the testing of a large number of promoter–gene combinations and wisely choosing the best. In addition, it might require a combination of two or more genes in multiple gene cascades to add their cumulative potential benefit. This might, for example, involve increasing N influx, increased flux from roots towards shoots, metabolic flux for assimilation into amino acids, and flux towards sequestration (e.g. storage into the vacuole). In addition to improving NUE through better N acquisition, focus on improving N remobilization would be quite fruitful through the efficient degradation of proteins and other organic N forms and transport during grain-filling. In addition, the components of NUE interact in multiple and complex ways with other metabolic pathways. Thereby, improving NUE is also dependent on other factors which affect N uptake, assimilation, and remobilization efficiency, such as C status, water, light, temperature, and soil type. Maintenance of optimum N and C homeostasis is required for optimum root and shoot development, photosynthetic rate, synthesis and mobilization of amino acids, organic acids, and lipids. Also, drought stress and N availability are linked for optimum yield, in that water use efficiency (yield per unit of transpiration) decreases when plant growth decreases with reduced N availability as photosynthesis is decreased more than transpiration (Cabrera-Bosquet et al., 2007). Combining multiple interlinked agronomic traits particularly growth under low N and drought tolerance is also important. In conclusion, there are many potential genetic approaches for the improvement of crop NUE and these require increased knowledge of how plants respond to different N regimes as well as to other environmental conditions.

SA is appeared one of the true stress responsive lant hormone which importantly contributes growth and development mechanism under adverse environmental conditions. Although, the biochemical mechanisms that mediate most of these responses remain unclear but NPR1 monomer functions as a co-activator of gene transcription which imparts to reveal the role SA in sustainable agriculture the identification and characterization of the SA receptor is probably the most anticipated discovery.

Secondary metabolism and transgenic plants has become a standard tool in plant stress biology. These technologies have mainly been utilized to model systems and have greatly enlarged the knowledge of mechanisms of tolerance. Phytohormones basically salicylic acid improves acclimation to abiotic stress by ion exclusion and/or compartmentalizetion, osmotic adjustments, reduction in lipid peroxidation, stress induced synthesis of protein kinases (SIPK), regulation of oxidative system and also enhanced antioxidant activity to combat the lethal oxidative events. Recently, it has been observed that reversible protein phosphorylation is likely to play a crucial role in mediating intracellular response to external stimuli in plants and suggest that this family of kinases may be involved in stress-signal transduction pathways in plants.

The molecular analysis of stress responses has arrived at a stage where research can facilitate a large collection of characterized genes. Identification and characterization of genetic loci for abiotic stress resistance could be an effective analytical tool. The use of novel approaches combining genetic, biochemical and molecular techniques should provide results in near future for food security.

5. References

Abreu ME, Munne´ -Bosch S. 2009. Salicylic acid deficiency in NahG transgenic lines and sid2 mutants increases seed yield in the annual plant Arabidopsis thaliana. Journal of Experimental Botany 60, 1261–1271.

Acar O, Türkan I, Ozdemir F (2001) Superoxide dismutase and peroxidase activities in drought sensitive and resistant barley (Hordeum vulgare L.) varieties. Acta Physiologiae Plantarum 23, 351-356

Adam Z, Adamska I, Nakabayashi K, et al. 2001. Chloroplast and mitochondrial proteases in Arabidopsis. A proposed nomenclature. Plant Physiology 125, 1912–1918.

Alscher RG, Erturk N, Heath LS (2002) Role of superoxide dismutases in controlling oxidative stress in plants. Journal of Experimental Botany 53, 1331-1341

Ananiev ED, Ananieva K, Todorov I. 2004. Effect of methyl ester of jasmonic acid, abscisic acid and benzyladenine on chlorophyll synthesis in excised cotyledons of Cucurbita pepo (Zucchini). Bulg J Plant Physiol 30: 51–63.

Asada K (1992) Production and scavenging of active oxygen in chloroplast In: Scandalios JG (Ed) Current Communications in Cell and Molecular Biology (Vol 5) Molecular Biology of Free Radicals Scavenging Systems, Cold Spring Harbour Laboratories Press, New York, pp 173-192

Badawi GH, Kawano N, Yamauchi Y, Shimada E, Sasaki R, Kubo A, Tanaka K (2004) Over-expression of ascorbate peroxidase in tobacco chloroplasts enhances the tolerance to salt stress and water deficit. Physiologiae Plantarum 121, 231-238

Baisak R, Rana D, Acharya PBB, Kar M (1994) Alterations in the activities of active oxygen scavenging enzymes of wheat leaves subjected to water stress. Plant Cell Physiology 35, 489-495

Baxter I. 2009. Ionomics: studying the social network of mineral nutrients. Current Opinion in Plant Biology 3: 381–386.

Borsani O, Valpuesta V, Botella MA (2001) Evidence for a role of salicylic acid in the oxidative damage generated by NaCl and osmotic stress in Arabidopsis Seedlings. Plant Physiology 126, 1024-1030

Borsani O., Valpuesta V., and Botella M. A. (2001), Evidence for a role of salicylic acid in the oxidative damage generated by NaCl and osmotic stress in Arabidopsis seedlings. Plant Physiol. 126, 1024D1030.

Bowler C, Van Camp W, Van Montagu M, Inzé D (1994) Superoxide dismutases in plants. Critical Research in Plant Sciences 13, 199-218

Bowler C, Van Montagu M, Inzé D (1992) Superoxide dismutase and stress tolerance. Annual Review of Plant Physiology and Plant Molecular Biology 43, 83-116

Bowling SA, Clarke JD, Liu Y, Klessig DF, Dong X. 1997. The cpr5 mutant of Arabidopsis expresses both NPR1-dependent and NPR1-independent resistance. The Plant Cell 9, 1573–1584.

Breusegem FV, Slooten L, Stassart JM, Moens T, Botterman J, Van Montagu M, Inzé D (1999) Overproduction of Arabidopsis thaliana FeSOD confers oxidative stress tolerance to transgenic maize. Plant Cell Physiology 40, 515-523

Broadley MR, Hammond JP, White PJ, Salt DE. 2010. An efficient procedure for normalizing ionomics data for Arabidopsis thaliana. New Phytologist 186: 270–274.

Brodersen P, Malinovsky FG, He´ maty K, Newman MA, Mundy J. 2005. The role of salicylic acid in the induction of cell death in Arabidopsis acd11. Plant Physiology 138, 1037–1045.

Campbell H.W. (1999): Nitrate reductase structure, function and regulation: Bridging the gap between biochemistry and physiology. Annual Review of Plant Physiology and Plant Molecular Biology, 50: 277–303.

Chen F, D'Auria JC, Tholl D, Ross JR, Gershenzon J, Noel JP, Pichersky E. 2003. An Arabidopsis thaliana gene for methylsalicylate biosynthesis, identified by a biochemical genomics approach, has a role in defense. The Plant Journal 36, 577–588.

Chen YM, Ferrar TS, Lohmeir-Vogel E, Morrice N, Mizuno Y, et al. 2006. The PII signal transduction protein of Arabidopsis thaliana forms an arginine-regulated complex with plastid N-acetyl glutamate kinase. J. Biol. Chem. 281:5726–33.

Chen Z, Zheng Z, Huang J, Lai Z, Fan B. 2009. Biosynthesis of salicylic acid in plants. Plant Signaling and Behavior 4, 493–496.

Chen Z,Watanabe T, Shinano T, Okazaki K, Osaki M. 2009. Rapid characterization of plant mutants with an altered ion-profile: a case study using Lotus japonicus. New Phytologist 181: 795–801.

Chisaki N, Horiguchi T. 1997. Responses of secondary metabolism in plants to nutrient deficiency. In: Ando T, Fujita K,Mae T, Matsumoto H, Mori S, Sekija J, editors. Plant nutrition for sustainable food production and environment. Dordrecht, The Netherlands: Kluwer academic publishers. pp. 341–345.

Conn S, Gilliham M. 2010. Comparative physiology of elemental distributions in plants. Annals of Botany 105: 1081–1102.

Coquoz JL, Buchala A, Me´ traux JP. 1998. The biosynthesis of salicylic acid in potato plants. Plant Physiology 117, 1095–1101.

Corpas FJ, Sandalio LM, del Río LA, Trelease RN (1998) Copper zinc superoxide dismutase is a constituent enzyme of the matrix of peroxisomes in cotyledons of oilseed plants. New Phytologist 138, 307- 314

Dat JF, Pellinen R, Cotte BVD, Langerbartels C, Kangasjarvi J, Inzé D, Van Breusegem F (2003) Changes in hydrogen peroxide homeostasis trigger an active cell death process in tobacco. Plant Journal 33, 621-632

Dean JV, Mills JD. 2004. Uptake of salicylic acid 2-O-b-D-glucose into soybean tonoplast vesicles by an ATP-binding cassette transporter-type mechanism. Physiologia Plantarum 120, 603–612.

Dean JV, Mohammed LA, Fitzpatrick T. 2005. The formation, vacuolar localization, and tonoplast transport of salicylic acid glucose conjugates in tobacco cell suspension cultures. Planta 221, 287–296.

Delorenzo L, Merchan F, Laporte P Thompson R Clarke J (2009). A novel plant leucine-rich repeat receptor kinase regulates the response of *Medicago trunculata roots* to salt stress. Plant Cell, 21:668-80.

Demidchik V, Essah PA, Tester M. 2004. Glutamate activates cation currents in the plasma membrane of Arabidopsis root cells. Planta 219:167–75.

Demiral T, Türkan I (2004) Does exogenous glycinebetaine affec antioxidative system of rice seedlings under NaCl treatment? Journal of Plant Physiology 161, 1089-1100

Demiral T, Türkan I (2005) Comparative lipid peroxidation, antioxidant defense systems and proline content in roots of two rice cultivars differing in salt tolerance. Environmental and Experimental Botany 53, 247-257

Ding G, Yang M, Hu Y, et al. 2010. Quantitative trait loci affecting seed mineral concentrations in Brassica napus grown with contrasting phosphorus supplies. Annals of Botany 105: 1221–1234.

Du L, Ali GS, Simons KA, Hou J, Yang T, Reddy ASN, Poovaiah BW. 2009. Ca2+/calmodulin regulates salicylic-acidmediated plant immunity. Nature 457, 1154–1158.

El Tayeb MA (2005) Response of barley grains to the interactive effect of salinity and salicylic acid. Plant Growth Regulation 45, 215-224

Eticha D, Zahn M, Bremer M, et al. 2010. Transcriptomic analysis reveals differential gene expression in response to aluminium in common bean (Phaseolus vulgaris) genotypes. Annals of Botany 105: 1119–1128.

Fariduddin Q, Hayat S, Ahmad A. 2003. Salicylic acid influences net photosynthetic rate, carboxylation efficiency, nitrate reductase activity, and seed yield in Brassica juncea. Photosynthetica 41: 281–284.

Fariduddin Q., Hayat S., Ahmad A. (2003): Salicylic acid influences net photosynthetic rate, carboxylation efficiency, nitrate reductase activity, and seed yield in Brassica juncea. Photosynthetica, 41: 281–284.

Ferrario-Mery S, Besin E, Pichon O, Meyer C, Hodges M. 2006. The regulatory PII protein controls arginine biosynthesis in Arabidopsis. FEBS Lett. 580:2015–20.

Filleur S,Walch-Liu P, Gan Y, Forde BG. 2005. Nitrate and glutamate sensing by plant roots. Biochem. Soc. Trans. 33:283–86.

Forde BG. 2002. Local and long-range signaling pathways regulating plant responses to nitrate. Annu. Rev. Plant Biol. 53:203–24.

Foyer CH, Souriau N, Perret S, Lelandais M, Kunert KJ, Pruvost C, Jouanin L (1995) Overexpression of glutathione reductase but not glutathione synthetase leads to increases in antioxidant capacity and resistance to photoinhibition in poplar trees. Plant Physiology 109, 1047-1057

Fu H, Dooner HK. 2002. Intraspecific violation of genetic colinearity and its implications in maize. Proc. Natl. Acad. Sci. USA 99:9573–78.

Fu¨hrs H, Behrens C, Gallien S, et al. 2010. Physiological and proteomic characterization of manganese sensitivity and tolerance in rice (Oryza sativa) in comparison with barley (Hordeum vulgare). Annals of Botany 105: 1129–1140.

Gallais A, Hirel B. 2004. An approach to the genetics of nitrogen use efficiency in maize. J. Exp. Bot. 55:295–306.

Gan Y, Filleur S, Rahman A, Gotensparre S, Forde BG. 2005. Nutritional regulation of ANR1and other root-expressed MADS-box genes in Arabidopsis thaliana. Planta 222:730–42

Gazaryan IG, Lagrimini LM (1996) Tobacco anionic peroxidase over expression in transgenic plants: aerobic oxidation of indole-3-acetic acid. Phytochemistry 42, 1271-1278

Glass ADM. 1974. Influence of phenolic acids upon ion uptake. J Exp Bot 25: 1104–1113.

Good AG, Shrawat AK, Muench DG. 2004. Can less yield more? Is reducing nutrient input into the environment compatible with maintaining crop production? Trends Plant Sci. 9:597–605.

Good AG, Shrawat AK, Muench DG. 2004. Can less yield more? Is reducing nutrient input into the environment compatible with maintaining crop production? Trends in Plant Science 9, 597–605.

Gunes A, Inal A, Alpaslan M, Eraslan F, Guneri Bagci E, Cicek N. 2007. Salicylic acid induded changes on some physiological parameters symptomatic for oxidative stress and mineral nutrition in maize (Zea mays L.) grown under salinity. Journal of Plant Physiology 164, 728–736.

Gutie´ rrez-Coronado MA, Trejo-Lo´ pez C, Larque´ -Saavedra A. 1998. Effects of salicylic acid on the growth of roots and shoots in soybean. Plant Physiology and Biochemistry 36, 563–565.

Hamilton E.W., Heckathorn S.A. (2001): Mitochondrial adaptations to NaCl. Complex I is protected by anti-oxidants and small heat shock proteins, whereas complex II is protected by proline and betaine. Plant Physiology, 126: 1266–1274.

Hayat Q., Hayat S., Ali B., Ahmad A. (2009): Auxin analogues and nitrogen metabolism, photosynthesis, and yield of chickpea. Journal of Plant Nutrition, 32: 1469–1485.

Hayat Q., Hayat S., Irfan M., Ahmad A. (2010): Effect of exogenous salicylic acid under changing environment: A review. Environmental and Experimental Botany, 68: 14–25.

Hayat S., Fariduddin Q., Ali B., Ahmad A. (2005): Effect of salicylic acid on growth and enzyme activities of wheat seedlings. Acta Agronomica Hungarica, 53: 433–437.

Hayat S., Hasan S.A., Fariduddin Q., Ahmad A. (2008): Growth of tomato (Lycopersicon esculentum) in response to salicylic acid under water stress. Journal of Plant Interactions, 3: 297–304.

Hemm MR, Rider SD, Ogas J, Murry DJ, Chapple C. 2004. Light induces phenylpropanoid metabolism in Arabidopsis roots. Plant J 38: 765–778.

Hernández JA, Jiménez A, Mullineaux P, Sevilla F (2000) Tolerance of pea (Pisum sativum L.) to long-term salt stress is associated with induction of antioxidant defenses. Plant Cell Environment 23, 853-862

Hernández JA, Olmos E, Corpas FJ, Sevilla F, del Río LA (1995) Salt-induced oxidative stress in chloroplasts of pea plants. Plant Science 105, 151-167

Hodge A. 2004. The plastic plant: root responses to heterogeneous supplies of nutrients. New Phytol. 162:9–24.

Holmström K.O., Somersalo S., Mandal A., Palva T.E., Welin B. (2000): Improved tolerance to salinity and low temperature in transgenic tobacco producing glycine betaine. Journal of Experimental Botany, 51: 177–185.

Hsieh MH, Lam HM, van de Loo FJ, Coruzzi G. 1998. A PII-like protein in Arabidopsis: putative role in nitrogen sensing. Proc. Natl. Acad. Sci. USA 95:13965–70.

Hsu SY, Kao CH (2003) Differential effect of sorbitol and polyethylene glycol on antioxidant enzymes in rice leaves. Plant Growth Regulation 39, 83-90

Hu Y, Bao F, Li J. 2000. Promotive effect of brassinosteroids on cell division involves a distinct CycD3-induction pathway in Arabidopsis. The Plant Journal 24, 693–701.

Jain A., Srivastava H.S. (1981): Effect of salicylic acid on nitrate reductase activity in maize seedlings. Physiologia Plantarum, 51: 339–342.

Kaiser WM, Huber SC. 2001. Post-translational regulation of nitrate reductase: mechanism, physiological relevance and environmental triggers. J. Exp. Bot. 52:1981–89.

Kang HG, Singh KB. 2000. Characterization of salicylic acidresponsive Arabidopsis Dof domain proteins: overexpression of OBP3 leads to growth defects. The Plant Journal 21, 329–339.

Kang JM, Turano FJ. 2003. The putative glutamate receptor 1.1 (AtGLR11) functions as a regulator of carbon and nitrogen metabolism in Arabidopsis thaliana. Proc. Natl. Acad. Sci. USA 100:6872–77.

Karley AJ, White PJ. 2009. Moving cationic minerals to edible tissues: potassium, magnesium, calcium. Current Opinion in Plant Biology 12: 291–298.

Kaydan D, Yagmur M, Okut N (2007) Effects of salicylic acid on the growth and some physiological characters in salt stressed wheat (Triticum aestivum L.). Tarim Bilimleri Dergisi 13, 114-119

Kichey T, Hirel B, Heumez E, Dubois F, Le Gouis J. 2007. In winter wheat (Triticum aestivum L.), post-anthesis nitrogen uptake and remobilisation to the grain correlates with agronomic traits and nitrogen physiological markers. Field Crops Research 102, 22–32.

Klessig DF, Malamy J (1994) The salicylic acid signal in plants. Plant Molecular Biology 26, 1439-1458

Klessig DF, Malamy J. 1994. The salicylic acid signal in plants. Plant Molecular Biology 26, 1439–1458.

Kobayashi T, Ogo Y, Aung MS, et al. 2010. The spatial expression and regulation of transcription factors IDEF1 and IDEF2. Annals of Botany 105: 1109–1118.

Koca H, Bor M, Ozdemir F, Türkan I (2007) The effect of salt stress on lipid peroxidation, antioxidative enzymes and proline content of sesame cultivars.Environmental and Experimental Botany 60, 344-351

Koch J. R., Creelman R. A., Eshita S. M., Seskar M., Mullet J. E., and Davis K. R. (2000), Ozone sensitivity in hybrid poplar correlates with insensitivity to both salicylic acid and jasmonic acid. The role of programmed cell death in lesion formation. Plant Physiol. 123, 487D496.

Kova´ cik J, Gru´ z J, Backor M, Strnad M, Repca'k M. 2009. Salicylic acid-induced changes to growth and phenolic metabolism in Matricaria chamomilla plants. Plant Cell Reports 28, 135–143.

Lacombe B, Becker D, Hedrich R, DeSalle R, Hollmann M, et al. 2001. The identity of plant glutamate receptors. Science 292:1486–87.

Lattanzio V, Cardinali A, Ruta C, Fortunato IM, Lattanzio VMT, Linsalata V, et al. 2009. Relationship of secondary metabolism to growth in oregano (Origanum vulgare L.) shoot cultures under nutritional stress. Environ Exp Bot 65: 54–62.

Lee HI, Leo´ n J, Raskin I. 1995. Biosynthesis and metabolism of salicylic acid. Proceedings of the National Academy of Sciences, USA 92, 4076–4079.

Lejay L, Tillard P, Lepetit M, Olive F, Filleur S, et al. 1999. Molecular and functional regulation of two NO3- uptake systems by N- and C-status of Arabidopsis plants. Plant J. 18:509–19.

Lindquist Y, Branden CL, Mathews FS, Lederer F (1991) Spinach glycolate oxidase and yeast flavocytochrome b2 are structurally homologous and evolutionarily related enzymes with distinctly different function and flavin mononucleotide binding. Journal of Biological Chemistry 266, 3198-3207

Little DY, Rao H, Oliva S, Daniel-Vedele F, Krapp A, Malamy JE. 2005. The putative high-affinity nitrate transporter NRT2.1 represses lateral root initiation in response to nutritional cues. Proc. Natl. Acad. Sci. USA 102:13693–98.

Liu J, Wu YH, Yan JJ, Liu YD, Shen FF. 2008. Protein degradation and nitrogen remobilization during leaf senescence. Journal of Plant Biology 51, 11–19.

Liu KH, Tsay YF. 2003. Switching between the two action modes of the dual-affinity nitrate transporter CHL1 by phosphorylation. EMBO J. 22:1005–13.

Madhusudhan R, Ishikawa T, Sawa Y, Shigeoka S, Shibata H (2003) Characterization of an ascorbate peroxidase in plastids of tobacco BY-2 cells. Physiologiae Plantarum 117, 550-557

Majeau N., Coleman J.R. (1994): Correlation of carbonic anhydrase and ribulose-1,5-bisphosphate carboxylase/oxygenase expression in pea. Plant Physiology, 104: 1393–1399.

Mäkelä P., Kärkkäinen J., Somersalo S. (2000): Effect of glycinbetaine on chloroplast ultrastructure, chlorophyll and protein content, and RuBPCO activities in tomato grown under drought or salinity. Biologia Plantarum, 43: 471–475.

Malamy J, Hennig J, Klessig DF. 1992. Temperature-dependent induction of salicylic acid and its conjugates during the resistance response to tobacco mosaic virus infection. The Plant Cell 4, 359–366.

Marschner H. 1995. Mineral nutrition of higher plants, 2nd edn. London: Academic Press.

Martin C, Bhatt K, Baumann K. 2001. Shaping in plant cells. Current Opinion in Plant Biology, 4: 540–549.

Masclaux-Daubresse C, Daniel-Vedele F, Dechorgnat J, Chardon F, Gaufichon L, Suzuki A. 2010. Nitrogen uptake, assimilation and remobilization in plants:

challenges for sustainable and productive agriculture. Annals of Botany 105: 1141–1158.

McClure JM 1997. The physiology of phenolic compounds. Recent Adv Phytochem 12: 525–556.

McCord JM (2000) The evolution of free radicals and oxidative stress. American Journal of Medicine 108, 652-659

Meneguzzo S, Navari-Izzo F, Izzo R (1999) Antioxidative responses of shoots and roots of wheat to increasing NaCl concentrations. Journal of Plant Physiology 155, 274-280

Mengel K, Kirkby EA, Kosegarten H, Appel T. 2001. Principles of plant nutrition. Dordrecht: Kluwer Academic.

Metraux J. P. (2002), Recent breakthroughs in the study of salicylic acid biosynthesis. Trends Plant Sci. 7, 332D334.

Metwally A., Finkemeier I., Georgi M., and Dietz K. J. (2003), Salicylic acid alleviates the cadmium toxicity in barley seedlings. Plant Physiol. 132, 272D281.

Miller AJ, Shen Q, Xu G. 2009. Freeways in the plant: transporters for N, P and S and their regulation. Current Opinion in Plant Biology 12: 284–290.

Mittler R, Blumwald E (2010) Genetic engineering for modern agriculture: challenges and perspectives. Ann. Review Plant Biol. 61:443-462.

Miwa K, Kamiya T, Fujiwara T. 2009. Homeostasis of the structurally important micronutrients, B and Si. Current Opinion in Plant Biology 12: 307–311.

Miyake K, Ito T, Senda M, Ishikawa R, Harada T, et al. 2003. Isolation of a subfamily of genes for R2R3-MYB transcription factors showing up-regulated expression under nitrogen nutrient-limited conditions. Plant Mol. Biol. 53:237–45.

Mohamed EA, Iwaki T, Munir I, Tamoi M, Shigeoka S, Wadano A (2003) Overexpression of bacterial catalase in tomato leaf chloroplasts enhances photooxidative stress tolerance. Plant Cell Environment 26, 2037-2046

Molina A., Bueno P., Marin M. C., Rodriguez-Rosales M. P., Belver A., Venema K., and Donaire J. P. (2002), Involvement of endogenous salicylic acid content, lipoxygenase and antioxidant enzyme activities in the response of tomato cell suspension cultures to NaCl. New Phytol. 156, 409D415.

Moorhead GB, Smith CS. 2003. Interpreting the plastid carbon, nitrogen, and energy status. A role for PII? Plant Physiol. 133:492–98.

Morcuende R, Krapp A, Hurry V, Stitt M. 1998. Sucrose feeding leads to increased rates of nitrate assimilation, increased rates of a-oxogluatarate synthesis, and increased synthesis of a wide spectrum of amino acids in tobacco leaves. Planta 206: 394–409.

Navarre DA, Mayo D. 2004. Differential characteristics of salicylic acid-mediated signaling in potato. Physiological and Molecular Plant Pathology 64, 179–188.

Nawrath C, Me´ traux JP. 1999. Salicylic acid induction-deficient mutants of Arabidopsis express PR-2 and PR-5 and accumulate high levels of camalexin after pathogen inoculation. The Plant Cell 11, 1393–1404.

Okabe K., Lindlar A., Tsuzuki M., Miyachi S. (1980): Effects of carbonic anhydrase on ribulose 1,5-bisphosphate carboxylase and oxygenase. FEBS Letters, 114: 142–144.

Overmyer K, Brosche M, Kangasjarvi J (2003) Reactive oxygen species and hormonal control of cell death. Trends in Plant Science 8, 335-342

Pal M., Horvath E., Janda T., Paldi E., and Szalai G. (2005), Cadmium stimulates the accumulation of salicylic acid and its putative precursors in maize (Zea mays) plants. Physiol. Plant. 125, 356D364.

Palenchar PM, Kouranov A, Lejay LV, Coruzzi GM. 2004. Genome-wide patterns of carbon and nitrogen regulation of gene expression validate the combined carbon and nitrogen (CN)-signaling hypothesis in plants. Genome Biol. 5:R91.

Park SW, Kaimoyo E, Kumar D, Mosher S, Klessig DF. 2007. Methyl salicylate is a critical mobile signal for plant systemic acquired resistance. Science 318, 113–116.

Puig S, Pen˜arrubia L. 2009. Placing metal micronutrients in context: transport and distribution in plants. Current Opinion in Plant Biology 12: 299–306.

Rai VK (2002) Role of amino acids in plant responses to stress. Biologiae Plantarum 45, 481-487

Rajjou L, Belghazi M, Huguet R, Robin C,Moreau A, Job C, et al. 2006. Proteomic investigation of the effect of salicylic acid on Arabidopsis seed germination and establishment of early defence mechanisms. Plant Physiol 141: 910–923.

Ramani B, Reeck T, Debez A, Stelzer R, Huchzermeyer B, Schmidt A (2006) Aster tripolium L. and Sesuvium portulacastrum L.: two halophytes, two strategies to survive in saline habitats. Plant Physiology and Biochemistry 44, 395-408

Raskin I (1992) Role of salicylic acid in plants. Annual Review of Plant Physiology and Plant Molecular Biology 43, 439-463

Raskin I, Skubatz H, Tang W, Meeuse BJD. 1990. Salicylic acid levels in thermogenic and non-thermogenic plants. Annals of Botany 66, 376–373.

Rate DN, Cuenca JV, Bowman GR, Guttman DS, Greenberg JT. 1999. The gain-of-function Arabidopsis acd6 mutant reveals novel regulation and function of the salicylic acid signaling pathway in controlling cell death, defenses and cell growth. The Plant Cell 11, 1695–1708.

Rate DN, Greenberg JT. 2001. The Arabidopsis aberrant growth and death2 mutant shows resistance to Pseudomonas syringae and reveals a role for NPR1 in suppressing hypersensitive cell death. The Plant Journal 27, 203–211.

Remans T, Nacry P, Pervent M, Girin T, Tillard P, et al. 2006. A central role for the nitrate transporter NRT2.1 in the integrated morphological and physiological responses of the root system to nitrogen limitation in Arabidopsis. Plant Physiol. 140:909–21.

Rich PR, Bonner WD Jr. (1978) The sites of superoxide anion generation in higher plant mitochondria. Archives of Biochemistry and Biophysics 188, 206-213

Riou-Khamlichi C, Huntley R, Jacqmard A, Murray JAH. 1999. Cytokinin activation of Arabidopsis cell division through a D-type cyclin. Science 283, 1541–1544.

Rizhsky L, Davletova S, Hongjian L, Mittler R (2004) The zinc finger protein Zat12 is required for cytosolic ascorbate peroxidase 1 expression during oxidative stress in Arabidopsis. Journal of Biological Chemistry 279, 11736-11743

Roxas VP, Smith Jr RK, Allen ER, Allen RD (1997) Overexpression of glutathione-S-transferase/glutathione peroxidase enhances the growth of transgenic tobacco seedlings during stress. Nature Biotechnology 15, 988-991

Roxas VP, Sundus AL, Garrett DK, Mahan JR, Allen RD (2000) Stress tolerance in transgenic tobacco seedlings that over express glutathione-Stransferase/glutathione peroxidase. Plant Cell Physiology 41, 1229-1234

Ru¨ schoff J, Wallinger S, Dietmaier W, Bocker T, Brockhoff G, Hofsta¨ dter F, Fishel R. 1998. Aspirin suppresses the mutator phenotype associated with hereditary nonpolyposis colorectal cancer by genetic selection. Proceedings of the National Academy of Sciences, USA 95, 11301–11306.

Rubio V, Bustos R, Irigoyen ML, Cardona-Lo'pez X, Rojas- Triana M, Paz-Ares J. 2009. Plant hormones and nutrient signalling. Plant Mol Biol 69: 361–373.

Sakhabutdinova AR, Fatkhutdinova DR, Shakirova FM (2004) Effect of salicylic acid on the activity of antioxidant enzymes in wheat under conditions of salination. Applied Biochemistry and Microbiology 40, 501-505

Salt DE, Baxter I, Lahner B. 2008. Ionomics and the study of the plant ionome. Annual Review of Plant Biology 59: 709–733.

Sanchez P. 2002. Soil fertility and hunger in Africa. Science 295:2019–20.

Sano H, Youssefian S. 1994. Light and nutritional regulation of transcripts encoding a wheat protein kinase homolog is mediated by cytokinins. Proc. Natl. Acad. Sci. USA 91:2582–86 .

Santner A, Calderon-Villalobos LIA, Estelle M. 2009. Plant hormones are versatile chemical regulators of plant growth. Nature Chemical Biology 5, 301–307.

Santner A, Estelle M. 2009. Recent advances and emerging trends in plant hormone signaling. Nature 459, 1071–1078.

Scheible WR, Gonzalez-Fontes A, Lauerer M, Muller-Rober B, Caboche M, Stitt M. 1997. Nitrate acts as a signal to induce organic acid metabolism and repress starch metabolism in tobacco. Plant Cell 9:783–98.

Scheible WR, Morcuende R, Czechowski T, Fritz C, Osuna D, et al. 2004. Genome-wide reprogramming of primary and secondary metabolism, protein synthesis, cellular growth processes, and the regulatory infrastructure of Arabidopsis in response to nitrogen. Plant Physiol. 136:2483–99.

Scott IM, Clarke SM, Wood JE, Mur LA. 2004. Salicylate accumulation inhibits growth at chilling temperature in Arabidopsis. Plant Physiology 135, 1040–1049.

Shah J. (2003), The salicylic acid loop in plant defense. Curr. Opin. Plant Biol. 6, 365D371.

Shah J. 2003. The salicylic acid loop in plant defense. Current Opinion in Plant Biology 6, 365–371.

Shakirova FM, Sakhabutdinova AR, Bezrukova MV, Fatkhutdinova RA, Fatkhutdinova DR (2003) Changes in the hormonal status of wheat seedlings induced by salicylic acid and salinity. Plant Science 164, 317-322

Shakirova FM, Sakhabutdinova AR, Bezrukova V, Fatkhutdinova RA, Fatkhutdinova DR. 2003. Changes in the hormonal status of wheat seedlings induced by salicylic acid and salinity. Plant Science 164, 317–322.

Shi Q, Bao Z, Zhu Z, Ying Q, Qian Q. 2006. Effects of different treatments of salicylic acid on heat tolerance, chlorophyll fluorescence and antioxidant enzyme activity in seedlings of Cucumis sativa L. Plant Growth Regul 48: 127–135.

Shulaev V, Silverman P, Raskin I. 1997. Airborne signalling by methyl salicylate in plant pathogen resistance. Nature 385, 718–721.

Singh A, Singh PK. 2008. Salicylic acid induced biochemical changes in cucumber cotyledons. I. J Agri Biochem 21: 35–38.

Singh B, Usha K. 2003. Salicylic acid induced physiological and biochemical changes in wheat seedlings under water stress. Plant Growth Regul 39: 137–141.

Singh B., Usha K. (2003): Salicylic acid induced physiological and biochemical changes in wheat seedlings under water stress. Plant Growth Regulation, 39: 137–141.

Singh PK, Koul KK, Tiwari SB, Kaul RK. 1997. Effect of cinnamate on nitrate reductase activity in isolated cucumber cotyledons. Plant Growth Regul 21: 203–206.

Song JT, Lu H, Greenberg JT. Stevens J, Senaratna T, Sivasithamparam K. 2006. Salicylic acid induces salinity tolerance in tomato (Lycopersicon esculentum cv. Roma): associated changes in gas exchange, water relations and membrane stabilisation. Plant Growth Regulation 49, 77–83.

Song JT. 2006. Induction of a salicylic acid glucosyltransferase, AtSGT1, is an early disease response in Arabidopsis thaliana. Molecules and Cells 22, 233–238.

Stitt M, Muller C, Matt P, Gibon Y, Carillo P, et al. 2002. Steps towards an integrated view of nitrogen metabolism. J. Exp. Bot. 53:959–70.

Stitt M, Muller C, Matt P, Gibon Y, Carillo P, Morcuende R, et al. 2002. Steps towards an integrated view of nitrogen metabolism. J Exp Bot 53: 959–970.

Sugiharto B, Suzuki I, Burnell JN, Sugiyama T. 1992. Glutamine induces the Ndependent accumulation of mRNAs encoding phosphoenolpyruvate carboxylase and carbonic anhydrase in detached maize leaf tissue. Plant Physiol. 100:2066–70.

Szalai G, Janda T (2009) Effect of salt stress on the salicylic acid synthesis in young maize (Zea mays L.) plants. Journal of Agronomy and Crop Science 195, 165-171

Szepesi A, Csiszar J, Bajkan Sz, Gemes K, Horvath F, Erdei L, Deer A, Simon LM, Tari I (2005) Role of salicylic aicd pre-treatment on the acclimation of tomato plants to salt- and osmotic stress. Acta Biologiae Szegediensis 49, 123-125

Tabuchi M, Abiko T, Yamaya T. 2007. Assimilation of ammonium ions and reutilization of nitrogen in rice (Oryza sativa L.). Journal of Experimental Botany 58, 2319–2327.

Takei K, Takahashi T, Sugiyama T, Yamaya T, Sakakibara H. 2002. Multiple routes communicating nitrogen availability from roots to shoots: a signal transduction pathway mediated by cytokinin. J. Exp. Bot. 53:971–77.

Taniguchi M, Kiba T, Sakakibara H, Ueguchi C, Mizuno T, Sugiyama T. 1998. Expression of Arabidopsis response regulator homologs is induced by cytokinins and nitrate. FEBS Lett. 429:259–62.

Tari I, Csiszar J, Szalai G, Horvath F, Pecsvaradi A, Kiss G, Szepesi A, Szabo M, Erdei L (2002) Acclimation of tomato plants to salinity stress after a salicylic acid pre-treatment. Acta Biologiae Szegediensis 46, 55-56

Tari I, Simon LM, Deer KA, Csiszar J, Bajkan Sz, Kis Gy, Szepesi A (2004) Influence of salicylic acid on salt stress acclimation of tomato plants: oxidative stress responses and osmotic adaptation. Acta Physiologiae Plantarum 26S, 237

Todd CD, Zeng P, Huete AM, Hoyos ME, Polacco JC. 2004. Transcripts of MYB-like genes respond to phosphorus and nitrogen deprivation in Arabidopsis. Planta 219:1003–9.

Torres MA, Dangl JL. 2005. Functions of the respiratory burst oxidase in biotic interactions, abiotic stress and development. Curr. Opin. Plant Biol. 8:397–403.

Traw MB, Bergelson J. 2003. Interactive effects of jasmonic acid, salicylic acid, and gibberellin on induction of trichomes in Arabidopsis. Plant Physiology 133, 1367–1375.

Türkan I, Bor M, Ozdemir F, Koca H (2005) Differential responses of lipid peroxidation and antioxidants in the leaves of drought-tolerant P. acutifolius Gray and drought-sensitive P. vulgaris L. subjected to polyethylene glycol mediated water stress. Plant Science 168, 223-231

Udvardi M.K., Day D.A. (1990): Ammonia (14C-methylamine) transport across the bacteroid and peribacteroid membranes of soybean root nodules. Plant Physiology, 94: 71–76.

Uzunova A.N., Popova L.P. (2000): Effect of salicylic acid on leaf anatomy and chloroplast ultrastructure of barley plants. Photosynthetica, 38: 243–250.

Uzunova AN, Popova LP. 2000. Effect of salicylic acid on leaf anatomy and chloroplast ultra structure of barley plants.Photosynthetica 38: 243–250.

Vandenabeele S, Van Der Kelen K, Dat J, Gadjev I, Boonefaes T, et al. 2003. A comprehensive analysis of hydrogen peroxide-induced gene expression in tobacco. Proc. Natl. Acad. Sci. USA 100:16113–18.

Vicente MRS, Plasencia J. 2011. Salicylic acid beyond defence: Its role in plant growth and development. J Exp Bot 62: 3321–3338.

Vlot AC, Dempsey DMA, Klessig DF. 2009. Salicylic acid, a multifaceted hormone to combat disease. Annual Review of Phytopathology 47, 177–206.

Walch-Liu P, Liu LH, Remans T,Tester M, Forde BG. 2006. Evidence that L-glutamate can act as an exogenous signal to modulate root growth and branching in Arabidopsis thaliana. Plant Cell Physiol. 47:1045–57.

Wang J, Zhang H, Allen RD (1999) Overexpression of an Arabidopsis peroxisomal ascorbate peroxidase gene in tobacco increases protection against oxidative stress. Plant Cell Physiology 40, 725-732

Wang R, Guegler K, LaBrie ST, Crawford NM. 2000. Genomic analysis of a nutrient response in Arabidopsis reveals diverse expression patterns and novel metabolic and potential regulatory genes induced by nitrate. Plant Cell 12:1491–509.

Wang R, Okamoto M, Xing X, Crawford NM. 2003. Microarray analysis of the nitrate response in Arabidopsis roots and shoots reveals over 1,000 rapidly responding genes and new linkages to glucose, trehalose-6-phosphate, iron, and sulfate metabolism. Plant Physiol. 132:556–67.

Wang YH, Garvin DF, Kochian LV. 2001. Nitrate-induced genes in tomato roots. Array analysis reveals novel genes that may play a role in nitrogen nutrition. Plant Physiol. 127:345–59.

Watanabe T, Broadley MR, Jansen S, et al. 2007. Evolutionary control of leaf element composition in plants. New Phytologist 174: 516–523.

White PJ, Bradshaw JE, Dale MFB, et al. 2009. Relationships between yield and mineral concentrations in potato tubers. HortScience 44: 6–11.

White PJ, Broadley MR. 2009. Biofortification of crops with seven mineral elements often lacking in human diets – iron, zinc, copper, calcium, magnesium, selenium and iodine. New Phytologist 182: 49–84.

White PJ, Hammond JP, King GJ, et al. 2010. Genetic analysis of potassium use efficiency in Brassica oleracea. Annals of Botany 105: 1199–1210.

Wildermuth MC, Dewdney J, Wu G, Ausubel FM. 2001. Isochorismate synthase is required to synthesize salicylic acid for plant defence. Nature 414, 562–565.

Wolters H, Ju¨ rgens G. 2009. Survival of the flexible: hormonal growth control and adaptation in plant development. Nature Reviews Genetics 10, 305–317.

Xia J, Zhao H, Liu W, Li L, He Y. 2009. Role of cytokinin and salicylic acid in plant growth at low temperatures. Plant Growth Regulation 57, 211–221.

Yalpani N, Silverman P, Wilson TMA, Kleier DA, Raskin I. 1991. Salicylic acid is a systemic signal and an inducer of pathogenesisrelated proteins in virus-infected tobacco. The Plant Cell 3, 809–818.

Yang Z. M., Wang J., Wang S. H., and Xu L. L. (2003), Salicylic acid-induced aluminum tolerance by modulation of citrate efflux from roots of Cassia tora L. Planta 217, 168D174.

Yoshimura K, Yabuta Y, Ishikawa T, Shigeoka S (2000) Expression of spinach ascorbate peroxidase isozymes in response to oxidative stress. Plant Physiology 123, 223-234

Yusuf M, Hasan SA, Ali B, Hayat S, Fariduddin Q, Ahmad A (2008) Effect of salicylic acid on salinity induced changes in Brassica juncea. Journal of Integrative Plant Biology 50, 1-4

Zhang J, Kirkham MB (1994) Drought stress induced changes in activities of superoxide dismutase, catalase and peroxidase in wheat species. Plant Cell Physiology 35, 785-791

Made in United States
Troutdale, OR
09/26/2026

54927601R00037